Verne the Sperm

Cynthia Anderson Sanchez, M.S.

Written and Illustrated by Cynthia Anderson Sanchez, M.S.

Copyright © 2016 Cynthia Anderson Sanchez, M.S

ISBN: 0692723706
ISBN-13: 9780692723708

Dedication

This book is dedicated to

My heroes:

...

To my mentor Dr. Sepehr Eskandari

for his unwavering support throughout the years and for providing me the opportunity to conduct the biological research that inspired me to write this book.

...

To my husband Joey Sanchez

for believing in me, in the small things and the big things, without fail.

Thanks!

ဆာၚ

FOREWORD

Verne the sperm was born in the mind of an aspiring, young-at-heart research scientist in the midst of performing experiments in the area of reproductive biology.

This book is intended to communicate a complex physiological process in a fun, entertaining and creative way. Here the complex subjects of spermatogenesis and the male reproductive system are written in the delightful rhyming format reminiscent of our childhood.

This book was written to teach, delight and to shed some light on how fun and exciting learning science can be!

ENJOY!
ဆာၚ

ACKNOWLEDGMENTS

Thank you to the members of CreateSpace and Kindle who assisted me in created this book.

A day in the life
of Verne the Sperm
is not so quite the life,
you'll learn.

There's peril and strife
and souls that yearn.
To find one wife's
the sole concern.

His world is strange
and his journey long
from the would-be dad
to the would-be mom.

In the human testes
his shape took form.
The little spermatozoon
was born!

Verne the sperm
is a simple foe.
He's covered with membrane
from head to toe.

On the tip of his head
is the acrosome
and inside his head
is the DNA's home.

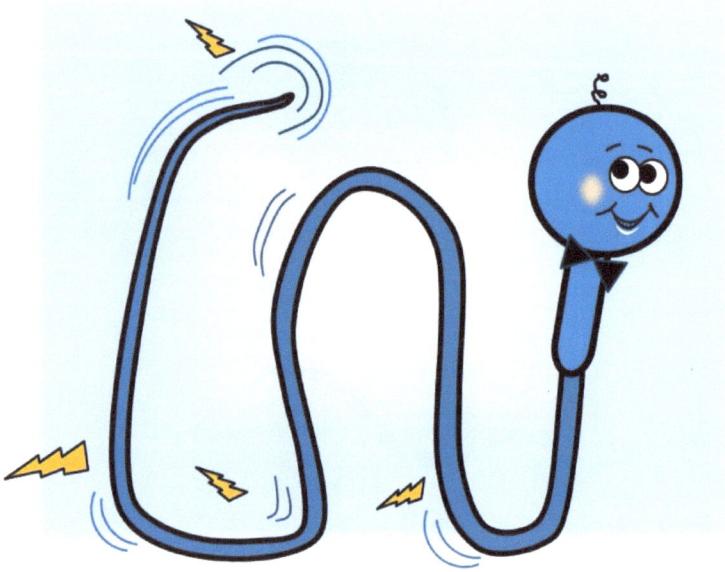

His body, the midpiece,
is short and stout
compared to his tail
without a doubt!

The tail, the flagellum,
I'd like to point out,
gives him mobility
to swim about.

Verne was a deep one
with deep thoughts, indeed!
He wondered and pondered
how he came to be.

He thought to himself,
"Who am I meant to be?"
Spermatogenesis!
This was the key!

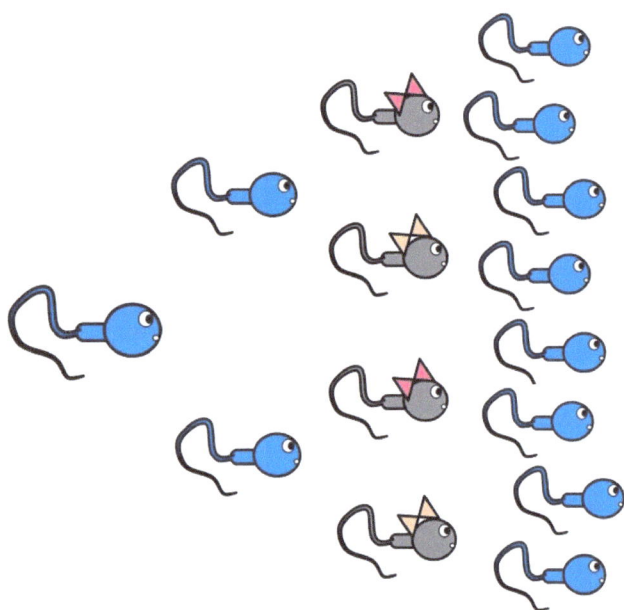

Information is here
in this book, on these pages,
on spermatogenesis
and how the sperm ages.

This process, you know,
has many stages,
each with its own
intermediate changes.

Spermatogenesis

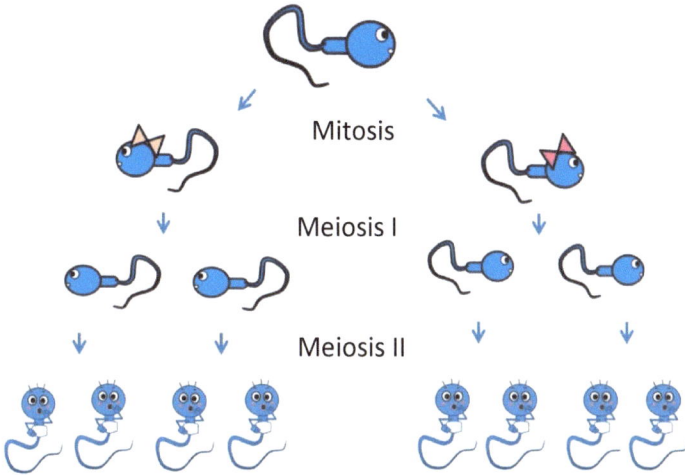

The first stage involves
the process mitosis.
The second stage involves
the process meiosis.

If I were not to be
such a good hostess,
I would not now try to
explain these in doses.

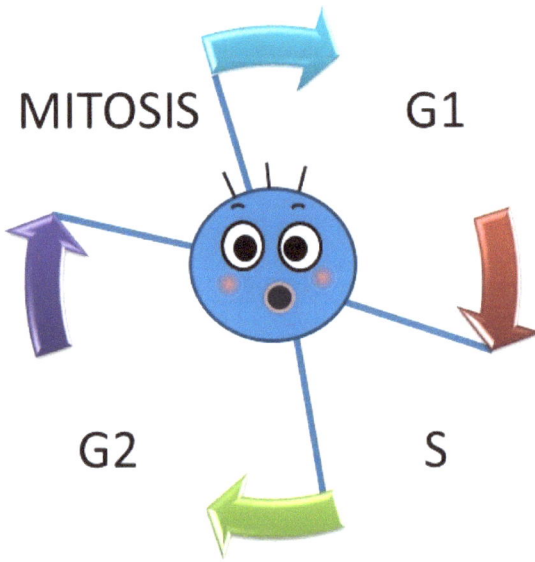

MITOSIS   G1

G2   S

First off, let me
introduce
how cells prepare
to reproduce.

The cell cycle is
how it starts.
Mitosis is just
one of those parts.

It started with Verne's
great grand-dad.
A spermatogium
that was named Brad.

Brad felt alone
and that made him sad.
So, he entered the cell cycle.
Then he was glad.

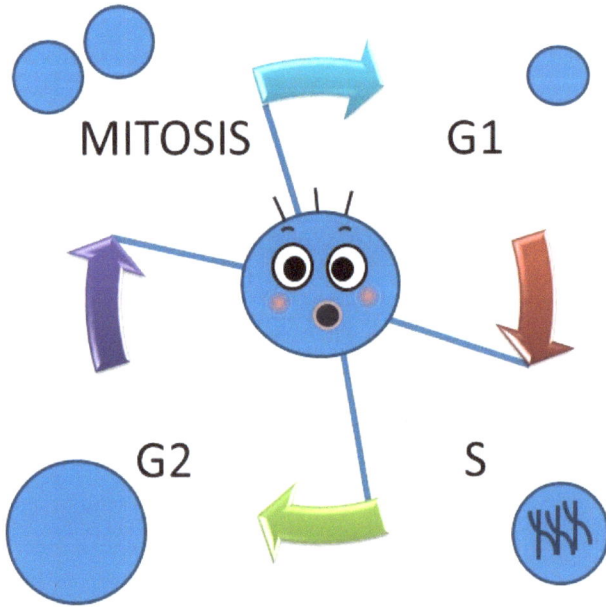

MITOSIS G1

G2 S

In G1, Brad
increases in size.
In S, his DNA
is synthesized.

In G2 Brad
gets bigger again.
Now his mitosis
is set to begin.

Mitosis

  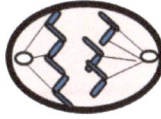 

Prophase   Metaphase   Anaphase

Telophase

In mitosis you
have prophase
and then metaphase,
like always,

and then anaphase
and telophase
when 2 cells
go their own ways.

Prophase

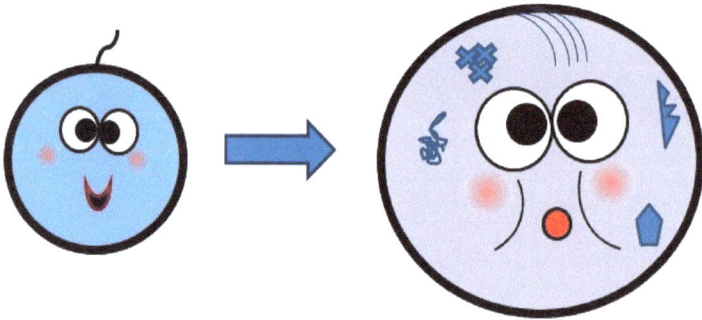

First we have prophase.
Brad, the parent cell,
at the start of mitosis
will start to swell.

He duplicates proteins
and each organelle.
His chromosomes form
from chromatin as well.

Metaphase

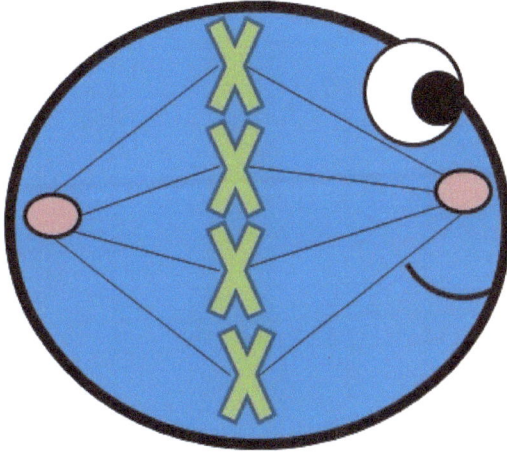

Grandfather Brad
has a certain prognosis
and enters the metaphase
stage of mitosis.

The chromosomes now
line up at the middle.
Their centromeres then
connect with the spindle.

The poles of the cell
contain centrosomes.
They pull on the spindles
toward each of their homes.

The spindles then pull
on each chromosome
like a tug-of-war game
of theirs all their own.

Anaphase

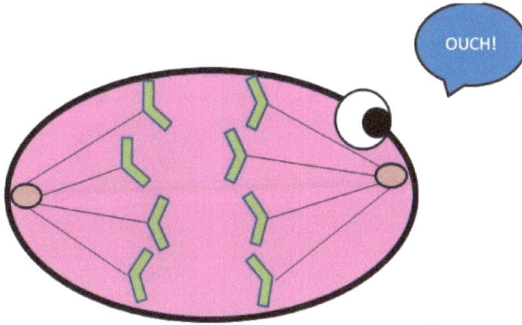

The sister chromatids
say their good-byes
and go their own ways
to live their own lives.

It is now when each
chromatid slides
during anaphase
toward opposite sides.

Telophase

Now in telophase
the membrane reforms.
Out of just one cell
now two will be born.

Chromosomes unwind
to form chromatin.
Cytokinesis brings
us to the end.

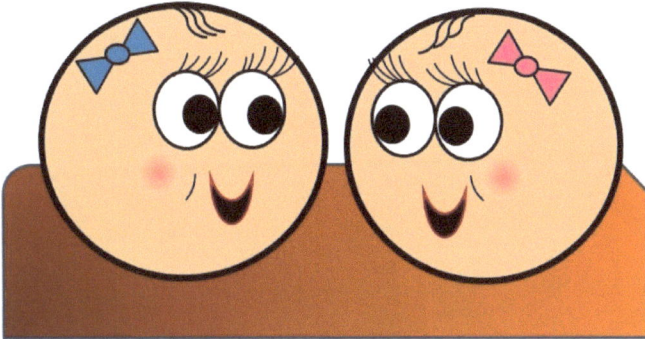

Each is a primary
spermatocyte.
With 46 chromosomes
they are alike.

They reside in the
basement membrane
of the seminiferous
tubules' domain.

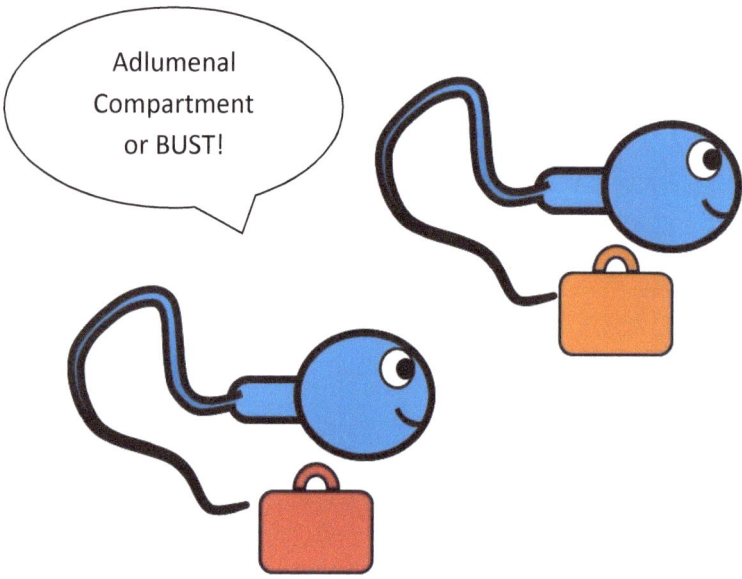

The spermatocytes
decided on movin'
from the basement
into the lumen.

They move into a
new apartment
in the adluminal
compartment.

Settle in quickly.
No time for fun.
They must soon begin
meiosis I.

And after that
they are not through.
They must then begin
meiosis II.

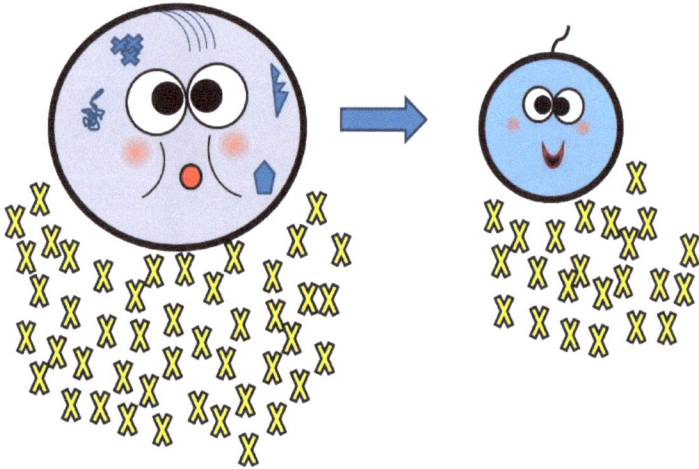

Meiosis is
similar indeed
to mitosis,
accept for one thing.

The 46 chromosomes
become 23.
There is a reduction to
haploid, you see.

The primary
spermatocytes
prepare themselves
for their new plights.

They knew it would be
an arduous one,
but now they enter
meiosis I.

Prophase I

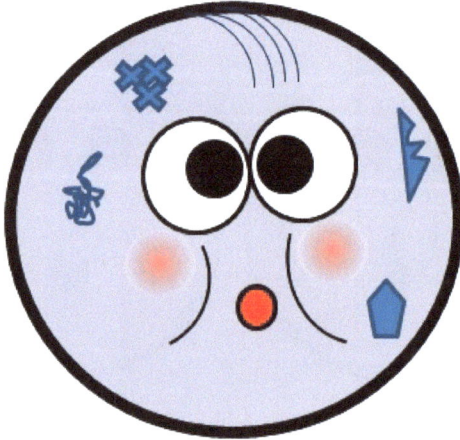

With prophase 1 it
sure does begin!
46 chromosomes
they have within.

The nuclear envelope
dissolves again.
The chromosomes
condense out of chromatin.

Metaphase I

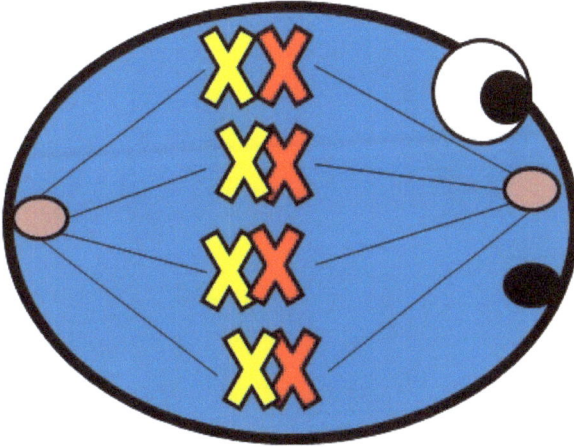

In metaphase 1
there is something new.
4 sister chromatids
instead of 2!

The chromatids line up
at the equator
forming bivalents.
More to come later!

Sister Chromatids

Something happens.
It's quite ethereal.
Here they swap
genetic material!

They do a dance.
They do a jig.
They each come out
a changed chromatid!

Are you interested?
I'll tell you moreover.
This 'swapping' event
is called crossing over.

In this way the daughter cells
won't be the same
Imagine if they were?
That would be lame!

Anaphase I

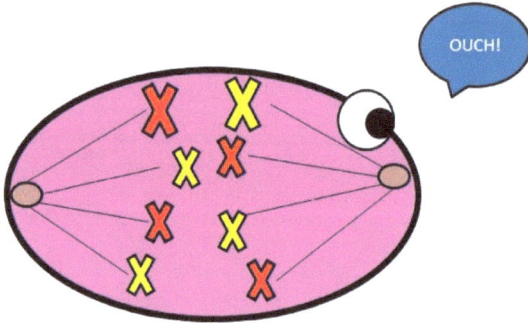

It's anaphase 1.
Let's celebrate!
Homologous chromosomes
now separate.

The whole chromosome's
the difference here.
Twice as much as
in mitosis, my dear.

Telophase I

In telophase 1
the spindle dissolves,
and nuclei reform
as the cell evolves.

Separation by
cytokinesis
brings forth 2 cells
as time increases.

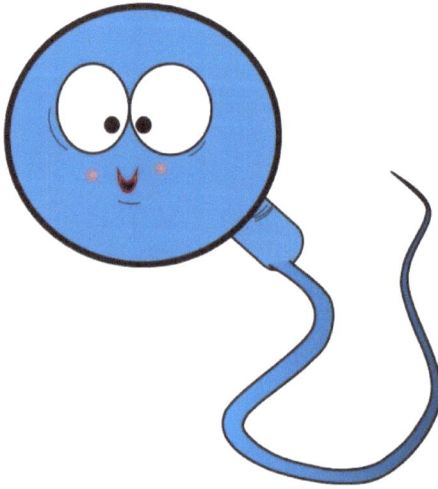

This is the first
of the arduous plight
to go from a spermato-
gonium right

into 2 primary
spermatocytes.
Yet, our dear Verne is still
nowhere in sight!

Ready for more science?
OK? Alright!

Meiosis II then
starts right away!
It's kind of a repeat.
This you could say.

There are some major
differences.
Let me tell you some
of these instances.

It starts with secondary
spermatocytes.
Compared to their parents
they are not alike!

Thanks to genetic
diversity,
I don't look like you
and you don't look like me.

Aren't you glad that no
person should suffer
to look like their dad
or look like their mother?

Mitosis

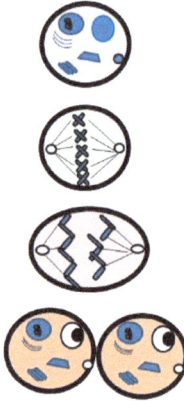

Meiosis I

Meiosis II

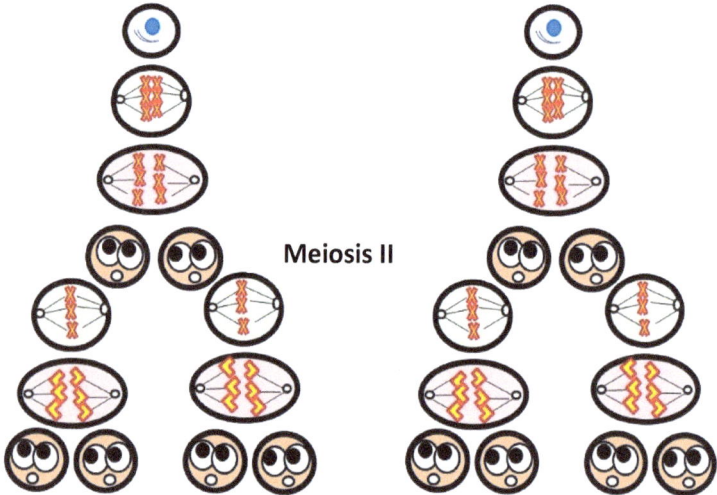

They start the phases
of meiosis II.
There's pro- and meta-
and telophase, too.

There is one difference
I should tell you,
The genetic material's
divided by 2.

So we went from Brad,
Verne's lovely grand-dad,
to the 2 primary
spermatocytes he had.

Both primary
spermatocytes
had 2 secondary
spermatocytes.

These spermatocytes
each have 2 kids.
When they are formed
they are called spermatids.

When they grow up
they are then called sperm,
including the special one
that we call Verne.

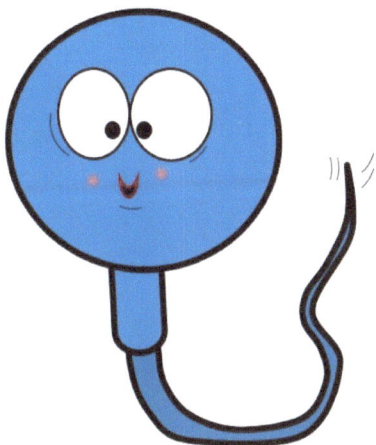

Verne was born in the
seminiferous tubules.
Verne looked all around
at all the geeks and ghouls.

He noticed his tail
was truncated.
It didn't do much.
This he sure hated!

.

Here we meet Verne
as just a kid.
An immature sperm
we call a spermatid.

He had a best friend!
Oh, yes he did!
A neighbor Sertoli cell
that he called Sid!

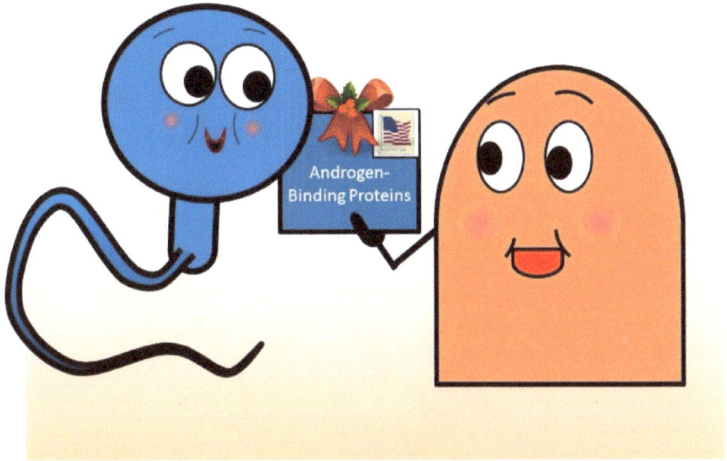

Sertoli cell Sid offered
lots of support.
He provided structure,
I'm glad to report!

Sid was anchored
to the membrane floor.
Androgen-binding
proteins he'd export.

This exportation,
I'll give explanation,
caused an infiltration
without hesitation.

Testosterone came in,
matured him and changed him.
He left to explore.
He was no longer caged in.

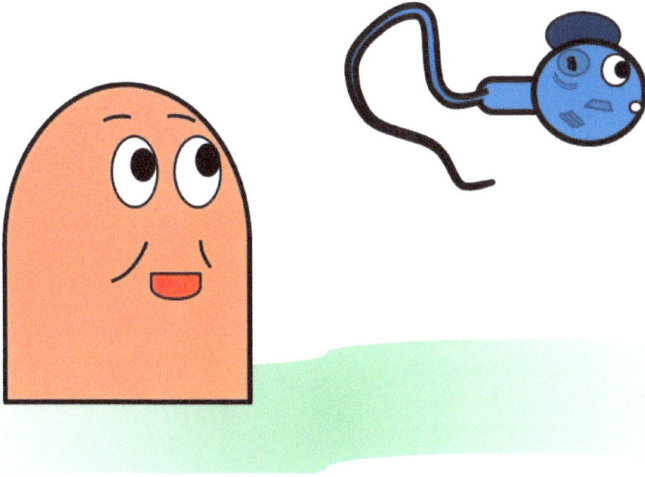

He left behind Sid,
who is now far below.
This is spermiation,
I thought you should know.

His nucleus became
a new acrosome.
He took off to venture
to find a new home.

Male Reproduction Tract

Sperm Passage

🟧	1.	Seminiferous Tubules
🟩	2.	Epididymis
⬛	3.	Van Deferens
⬜	4.	Ejaculatory Duct
🟥	5.	Urethra
🟨	6.	Penis

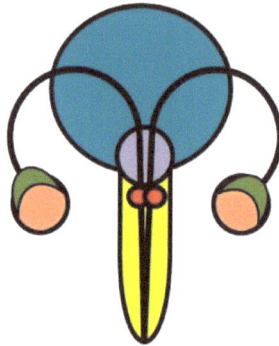

He started his journey
into the lumen.
Through seminiferous
tubules he was groovin'.

To the epididymis
he was still movin'
surrounded by proteins
like human albumin.

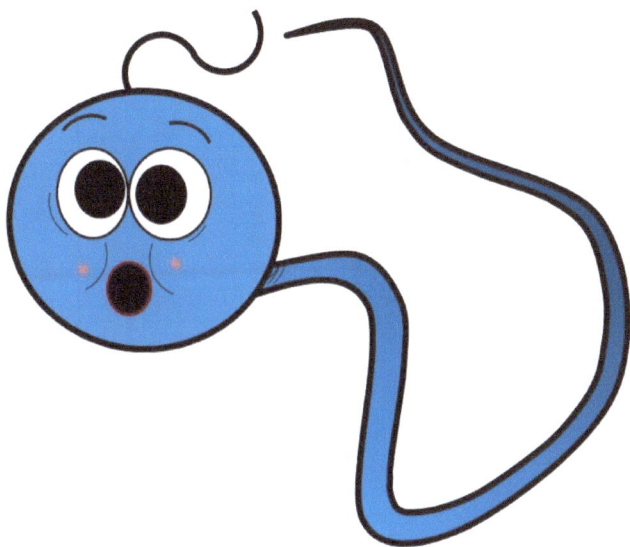

The journey took Verne
a couple of day days.
He swam through that
tubule maze.

He stopped to rest,
since rest always pays,
until a disruption
woke him from his daze.

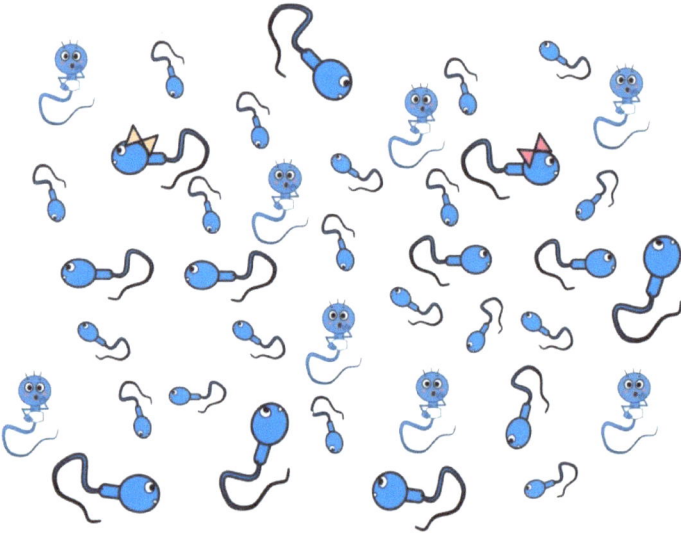

There was now traffic!.
It was hit and miss!
He fought his way to
the epididymis.

It was a whole city
with millions of sperm.
A world to explore.
A new world to learn.

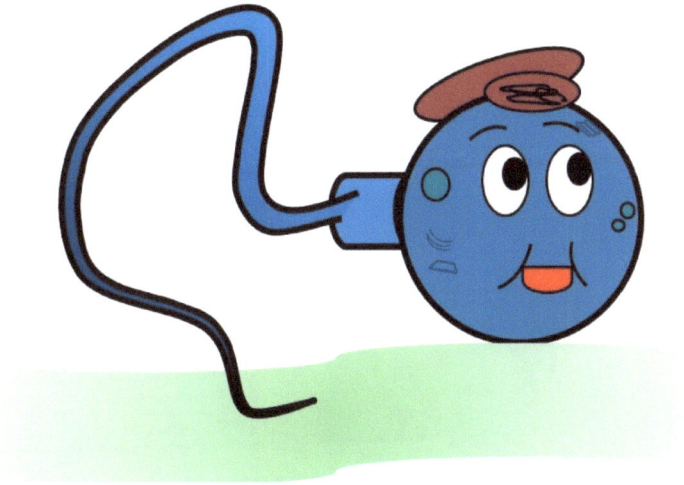

Here in the
epididymis
each sperm
has a pilgrimage.

They travel from the
head to the tail.
They mature on their
way without fail.

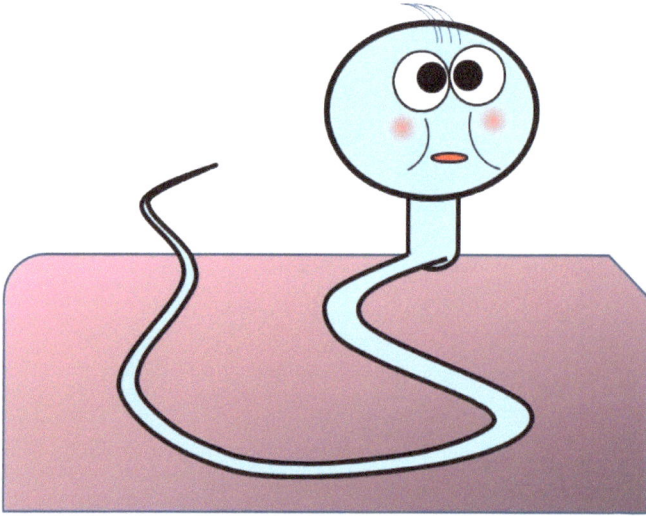

Verne took this trip.
It took him 2 weeks.
At this point,
a vacation he needs!

He relaxes into
complacency,
until a rumble woke him
from his sleep.

The sound was loud
and the lumen was shook!
From the epididymis
our Verne was took.

His journey seemed long
like there was no end.
He went into the
vas deferens.

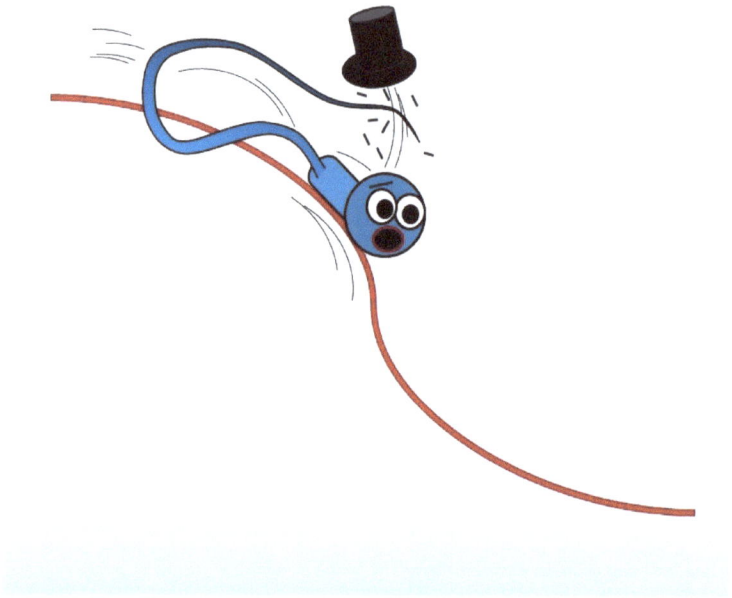

The trip was like
a water slide.
Verne thought he
was gonna die!

He landed in the
seminal vesicles.
His power here was
infinitesimal.

From the vesicles
to the prostate gland,
he tried to gain
the upper hand.

The fluid around
stuck to his skin.
as a protective
barrier for him.

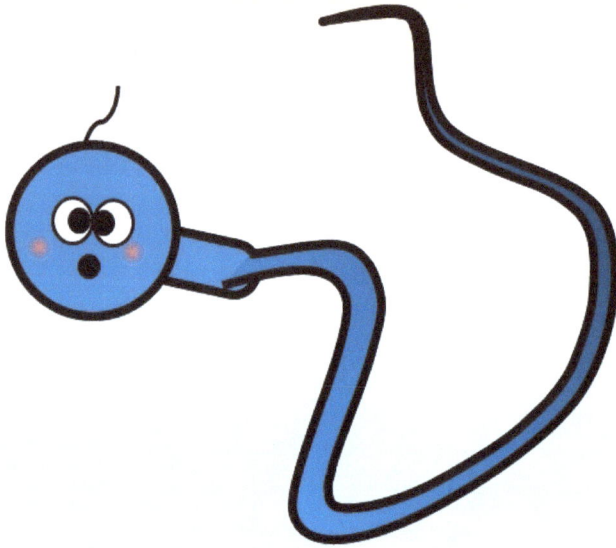

Are you ready
for the sequel?
Into the gland,
the bulborethral!

He journey seemed long
and somewhat lethal.
Can you believe this
goes on in people?

From this gland our
Verne was pushed.
He found himself
simply ambushed.

Now down the urethra
our little Verne flew!
He noticed a bad smell.
He said, "Pee-eew!".

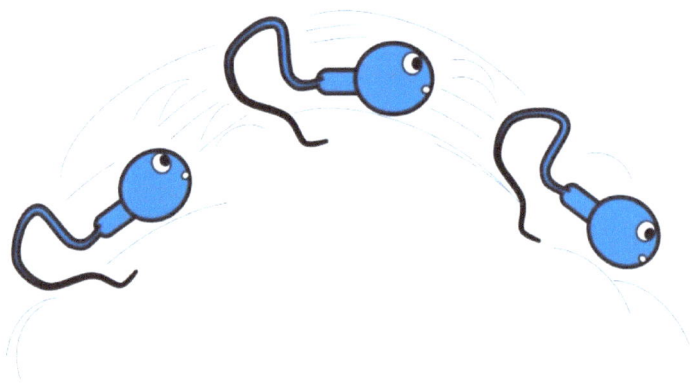

Next he flew out
of the urethra.
He shouted out loud,
"Ah! Eureka!".

It all went too fast
to think it through.
I would have been scared!
You would have been, too!

The sperm fell to the
ground with a 'splat!'.
He was now free
as a matter of fact.

He began to gather himself
from the ground.
He took just a moment
to look around.

He adjusted so well
to the world. I commend!
He was glad his adventure
had come to an end.

Yet somehow he knew it,
somewhere deep within,
another adventure
was soon to begin!

So, this is the story
about our dear Verne.
It is my hope that somehow
you have learned

the pathway we take
may have twists and turns,
but it's just an adventure
and joy will return.

The End

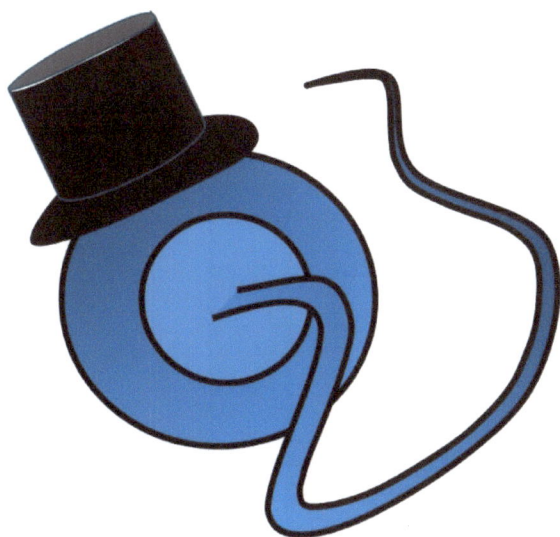

ဆာ

ABOUT THE AUTHOR

Cynthia Anderson Sanchez, M.S. has combined her love for the sciences and her love for teaching in this novel book on spermatogenesis and the sperm's journey through the male reproductive tract. She hopes that the readers will enjoy the science presented in this unique, fun and create manner in which this book was written.

Cynthia has earned a Bachelor of Science degree in Neuroscience, a Master of Arts degree in Education and a Master of Science degree in Biological Sciences. She continues teaching and her involvement in reproductive biology and neuroscience research.

ENJOY!

ဆာ

www.ingramcontent.com/pod-product-compliance
Lightning Source LLC
Chambersburg PA
CBHW041715200326
41519CB00001B/169